Just Fix This!

I0814005

BY KIM THOMPSON

A Little Honey Book

Tips for Teachers and Caregivers

This book supports early readers as they decode words to learn facts and gain knowledge about the world.

Before reading, make sure students understand the sound-spelling correspondences shown below as well as the high-frequency words shown on the next page. Introduce the vocabulary words.

During reading, provide feedback and encouragement as students sound out decodable words by blending individual sounds.

After reading, talk about and write about the topic. Share the information on page 16 to help students learn more.

Letters and Sounds

New:

Sound	Spelling
/j/	j
/v/	v
/ks/	x

Review:

Sound	Spelling
short a	a
/b/	b
/d/	d
short e	e
/f/	f
/g/	g
/h/	h
short i	i
/k/	c, ck, k
/l/	l
/m/	m
/n/	n
short o	o
/p/	p

/r/	r
/s/	s
/t/	t
short u	u

Decodable Words

an, ax, box, dog, fix, gets, got, helps, in, is, it, jack, jam, jet, job, jogs, jug, junk, just, juts, mix, must, Nan, next, on, rag, rev, rug, six, up, vac, van, vat, visit

High-Frequency Words

New: all, day, do, from, need, or, will

Review: a, for, live, made, out, the, there, this, was, what, you

Vocabulary Words

car

mower

states

storm

tool

A **tool** helps.

It is made just for the job.

There was a **storm**.
This juts out.

Do you need an ax or a **mower**?

An ax will fix it!

The dog jogs in.

Junk gets on the rug.

Do you need a rag or a vac?

A vac will do the job!

You will visit Nan. You live six **states** from Nan.

Do you need a van or a jet?

A jet gets you there
the next day!

You must mix. Mix all this!

Do you need a jug or a vat?

A vat will do the job!

The **car** got in a jam.
What will help it rev up?

What do you need from the box?

A jack will help fix it!

What do you need for this?

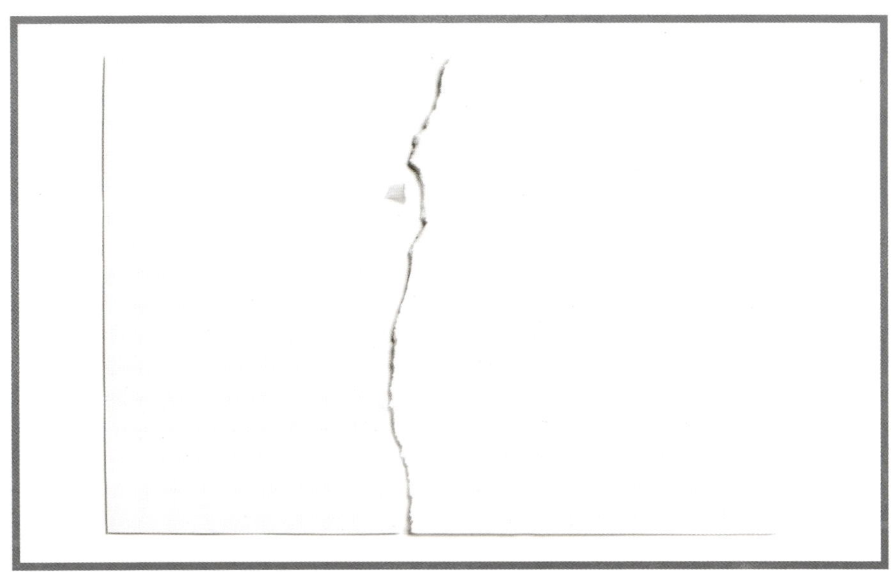

What tool will fix it?

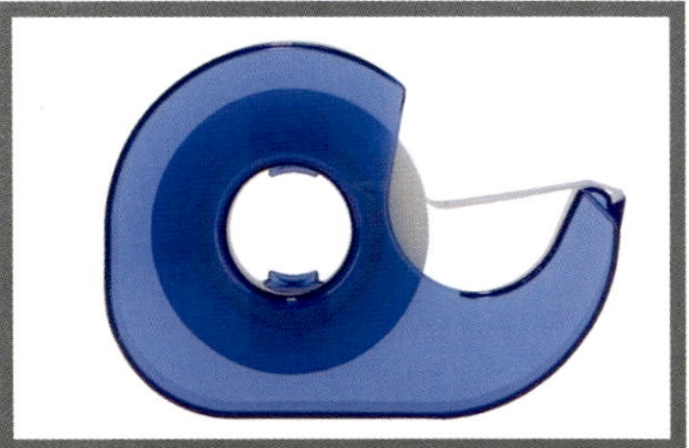

Build Background Knowledge

Tools help solve problems. Think about each tool and the problem it solves: a rake, a zipper, a pair of earbuds, a motorcycle, a laptop. People invent new tools to solve new problems. They change old tools to make them work better. A tool's shape affects the way it works. Think of a garden shovel, a sandbox shovel, a spoon, and an excavator. How does the shape of each tool help it dig different materials?

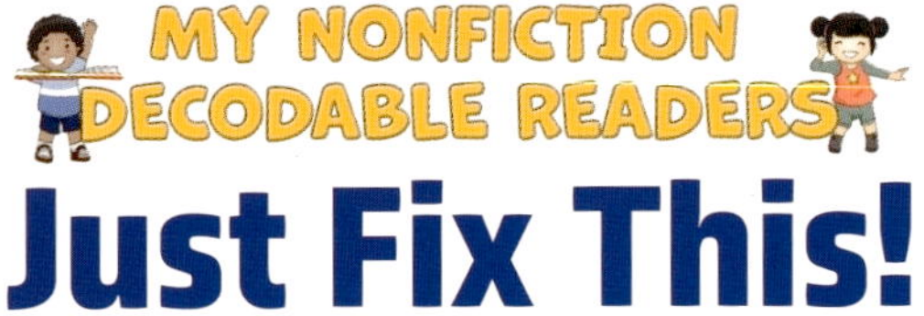

Just Fix This!

Written by: Kim Thompson
Designed by: Rhea Magaro
Series Development: James Earley
Educational Consultant: Marie Lemke, M.Ed.

Photographs: All images from Shutterstock

Crabtree Publishing

crabtreebooks.com 800-387-7650

Printed in China/012024/FE20231222

Published in Canada
Crabtree Publishing
616 Welland Ave.
St. Catharines, Ontario
L2M 5V6

Published in the United States
Crabtree Publishing
347 Fifth Ave
Suite 1402-145
New York, NY 10016

Library and Archives Canada Cataloguing in Publication
Available at Library and Archives Canada

Library of Congress Cataloging-in-Publication Data
Available at the Library of Congress

Hardcover: 978-1-0398-4434-6
Paperback: 978-1-0398-4515-2
Ebook (pdf): 978-1-0398-4592-3
Epub: 978-1-0398-4662-3
Read-Along: 978-1-0398-4732-3
Audio: 978-1-0398-4802-3